God's Cool Creation:

Tricky Trees!

God's Cool Creation: Tricky Trees (Book Series, Book 2)

First paperback edition: July 2022
ISBN: 979-8-218-03553-2
LCCN: 2022910330

Written and illustrated by Mary Ann Winslow, PhD
Contribution by Benjamin Winslow

Cool Creation Press
Prescott, Arizona

maryann@coolcreationpress.com
maryannwinslow.blogspot.com

Disclaimer: This book contains rudimentary science information that is intended to help the young reader understand basic principles. It is understood that some specific details have been omitted. For example, osmosis is not discussed when referring to water uptake in tree roots, nor is polarity. Phototropism, gravitropism, desert plants, and other topics and concepts are only given a cursory explanation.

This book is dedicated to
DAWSON♥

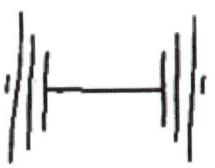

God made over 7,000 kinds of trees! There are over 100 kinds of pine trees alone.

The Bristlecone pine can live up to 5,000 years! That means that the same tree that was alive in Jesus' day is alive today!

Like us, trees and plants need water to live. And God's water is one of a kind! Water droplets like to stick together (cohesion - coh hee zhun) like best buddies! And they like to stick to other surfaces (adhesion - add hee zhun), too!

But how does water get from way down in the tree roots to the very tops of trees? Well, trees are busy inside!

There are tiny tubes, like super thin straws, throughout the tree called xylems (zy lumz). The water buddies use these straws (xylems) to travel from the soil and into the roots.

They use adhesion to climb up the straw (xylem), which goes through the trunk and branches, and reaches into the leaves!

Trees like the sequoia can be 250 feet high! That's 5 or 6 telephone poles stacked up on top of one another! That's a long way to climb for our water buddies!

So just how do the water buddies get pulled all the way up the straws (xylems)? It's something special in God's design called transpiration (tranz pur ay shun).

Water exits the leaf into the air through tiny holes under the leaf called stomata (stoh muh tuh).

This loss of water - evaporation (ee vapp orr ay shun) - causes the air to pull up the water below - all the way up! As the Sun beats down and it gets warmer outside, the transpiration increases!

What about desert plants and trees? It's sooo hot and dry in the desert! Do they work the same way?

Desert plants and trees have their own very tricky ways to get water!

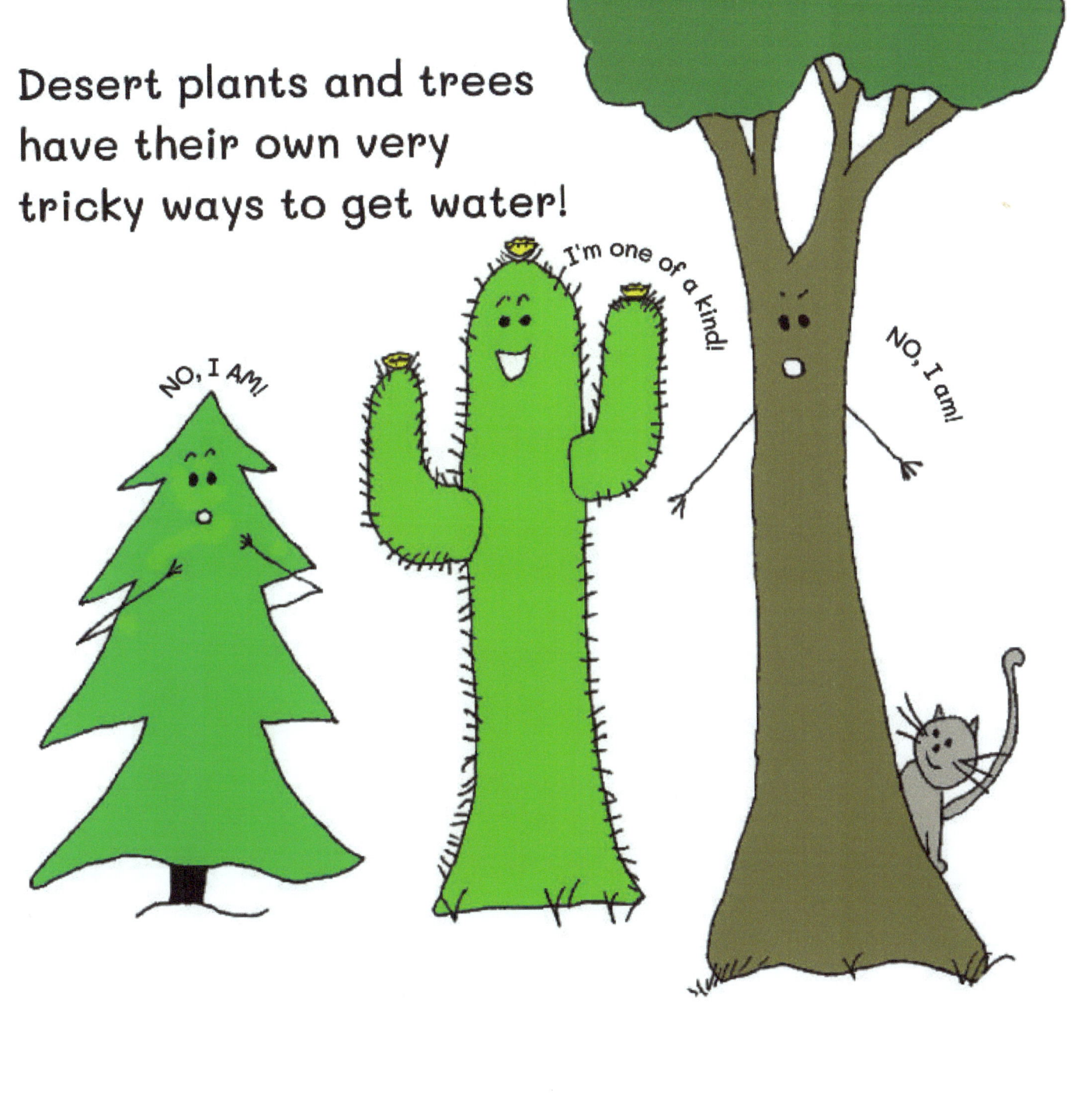

Some have some sharp spines to shade the plant and keep it cool, like the ocotillo (oh kuh tee uh).

Some have thorny spikes so animals won't want to eat them, like the prickly pear cactus.

Some have roots that can reach down 100 feet and drink up the water deep in the ground, like the mesquite (muh skeet).

Some have very shallow roots that really spread out, drinking up all the water around it, like the barrel cactus.

They can grow 10 feet tall!

And live 100 years. Don't touch!

Some plants spit out poison to kill any plants around it so that they can drink up all the water, like the creosote (cree oh soht) bush.

Most desert plants are thick and waxy. This protects the plant from drying out.

Some have few or no leaves, so no stomata.
That means that less water exits the plant, like
the saguaro cactus.

Many desert plants close
their stomata during the
day so they don't lose
water when it's the hottest.

Time to open up!

It's called phototropism (foh toh troh pizm)! Trees have a chemical called auxin (awks inn) that gathers to the shady side of the tree.

This causes the shady side of the tree to grow, causing the *sunny* side to curve up to the light! And plants need light to grow.

Bragger!

I'm needed for all life!

No she's not!

More tricks by trees! Did you ever wonder how trees can grow on rocks? Well, small amounts of soil and old plant parts gather in cracks of rocks that have been blown in by the wind.

Then seeds are blown in by the wind, too, or brought by birds and other animals. Before you know it - a tree!

God has created all this just for us!

Thank you, Lord!

Don't forget to read God's Cool Creation: Epic Earth!
More books coming soon!
God's Cool Creation:
Whimsical Water
Vivid Volcanoes
Dashing Desert
Arresting Arctic
Sensational Sea Mammals
Sassy Cells
Lovable Leaves
Bedazzling Birds

About the author: Mary Ann Winslow is a former science teacher, university instructor (University of Arizona, University of Wyoming, Texas A&M University), SIU Saluki, U of A Wildcat (Bear Down!), mom, grandma, and most importantly, Jesus follower. She currently resides in Prescott, Arizona, with her husband, Kent.